SNOW-FLAKES:

A CHAPTER FROM THE

BOOK OF NATURE.

HAST THOU ENTERED INTO THE TREASURES OF THE SNOW?
Job xxxviii. 22.

"Acquaint thyself with God, if thou wouldst taste
His works. Admitted once to his embrace,
Thou shalt perceive that thou wast blind before;
Thine eye shall be instructed; and thine heart,
Made pure, shall relish with divine delight,
Till then unfelt, what hands divine have wrought."
COWPER.

PREFACE

BRIEF article on Snow-flakes, in one of the periodicals published by the American Tract Society in the winter of 1862–3, accompanied by a cut exhibiting some of their forms, elicited from its readers many expressions of interest, and suggested the preparation of a book on this curious but generally little-known subject.

The beautiful forms of many of the snow-crystals were observed and sketched more than a century ago. The *Transactions of the Royal Society of London*, for 1755, contain representations of ninety-one varieties, with descriptions by Dr. Nettis. Captain William Scoresby, the eminent English navigator, has given, in his *Arctic Regions*, drawings of ninety-six varieties. More recently, numerous specimens, with accompanying descriptions, have been given to the public by James Glaisher, Esq., of Lewisham, England. It is from these sources, chiefly, that the figures here exhibited have been derived.

It has been no part of our design, in this work, to enter into any scientific statements concerning the snow-flakes, or the laws of their formation. A brief general description of them is all that has been attempted. Yet the reader should not, from this, infer that there is any question respecting the truthfulness of the sketches. The drawings were originally made with scientific precision, and have been carefully copied. A few simple figures at the top of page 11 are designed to show the primary geometrical *forms* under which the snow-vapor crystallizes. With that exception, they are all representations of individual crystals, actually observed and sketched with the aid of the microscope.

It is proper to add, however, that these representations are highly magnified, especially those on the last two or three plates. The real size of the crystals observed by Scoresby varied from one thirty-fifth to one-third of an inch in diameter. Dr. Nettis remarks: "The natural size of most of the shining, quadrangular particles, and of the little stars of snow, as well the simple as the more compound ones, does not exceed the twentieth part of an inch." The dimensions as well as form of the crystals seem to depend upon the amount of vapor in the atmosphere, the temperature, and other circumstances not easy to specify.

We may be permitted to express the hope that many of our readers will examine for themselves these beautiful productions of nature. In our own climate, the "treasures of the snow" are open to all who choose to explore them; and there can scarcely be an amusement more entertaining, and at the same time instructive, than that of observing and sketching these delicate crystals. No expensive or complicated apparatus is needed for this purpose. A good microscope is the chief requisite; besides which, a pair of dividers and a rule will be sufficient. We subjoin a statement from Mr. Glaisher of his own mode of making his examinations.

"For the information of those who would carefully observe snow-crystals, I may remark that my own plan of procedure is to expose a thick surface of plate-glass on the outer side of the window, resting on the ledge. Seated within the room, I am enabled, with comparative comfort, and at my leisure, to make my drawings and record my observations, the accuracy of which I am able to verify to my satisfaction, as the crystal received upon the cold surface of the glass, itself several degrees below freezing, remains a sufficient length of time for the requirements of the observer. In many cases, it becomes frozen to the glass, and is thus secured from the influence of the wind, which not unfrequently snatches away some most intricate form from the desiring eye of the observer."

If this work shall be the means of introducing any of our readers to the knowledge of this interesting department of the Creator's works, and eliciting those sentiments of admiration and reverence which his wonder-working power should inspire in every beholder, it will not have been issued in vain.

I. P. W.

BOSTON, 1863.

CONTENTS

CONTENTS.

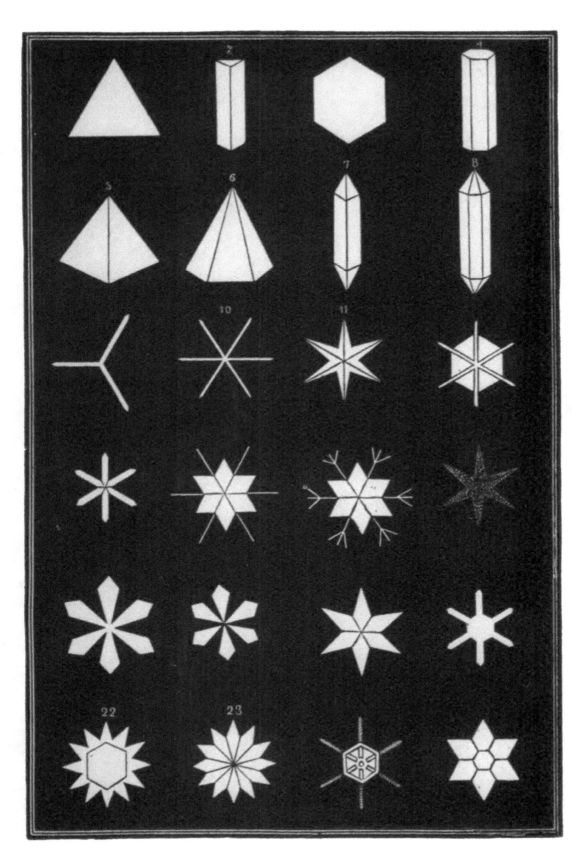

Snow Structure

He saith to the Snow, Be thou on the face of the Earth.— Job 37 : 6.

—————— ⁘ ——————

When the watery vapors in the atmosphere are in sufficient quantities to be precipitated to the earth, and at the same time their temperature is at or below the freezing point, their particles unite, but not as fluid drops. In approaching each other they arrange themselves in regular figures, called *crystals*. The various forms of these may be grouped into three general classes.

1. *Prismatic*, having three or six sides, usually the latter (page 11, figs. 2, 4). Scoresby compares the finest specimens of these to "white hairs cut into lengths not exceeding a quarter of an inch."

2. *Pyramidal*, either triangular or hexagonal (figs. 5, 6). They are exceedingly small, being only one-thirtieth of an inch in hight.

3. *Lamellar*, consisting of thin and flat plates, some of them stelliform, having six points radiating from a center (fig. 11), and some hexagonal (page 21, fig. 1). Both these species are in infinite abundance, and of all sizes, from the smallest speck to one-third of an inch in diameter.

These three leading forms are endlessly combined, and give rise to innumerable varieties, from the simplest to the most complex. Pyramids are mounted on prisms, at one or both ends (page 11, figs. 7, 8); prisms are united in one star-like figure, like spokes of a wheel (fig. 10), and both are joined with plates in all conceivable forms of beauty and diversity. The specimens shown throughout our series of engravings illustrate these. The plates themselves are complex, showing within their outer boundaries white lines, which divide them into triangles, stars, hexagons, and other regular figures. Some plates are transparent, others opaque (page 21, fig. 13).

When the prisms are combined with plates, it is generally in the same plane, but sometimes the former are set perpendicularly to the surfaces of the latter (page 29, figs. 18, 19, 20, 21). These singular figures resemble a wheel with its axle. Scoresby says that on one occasion, snow of this kind fell upon the deck of his ship to the depth of three or four inches!

In some instances the central plate has little prisms or spines projecting from it like hairs, on one or both sides, at an angle of sixty degrees. Sometimes, instead of a plate, the central part is a little rough mass like a hailstone, bristling with spines, somewhat resembling a chestnut-bur.

Much attention has been given to the meteorological conditions of the atmosphere during the fall of snow, to ascertain in what circumstances the different varieties of crystals are produced. Nothing very definite, however, is discoverable in this respect. The general facts are thus summed up by Mr. Scoresby: "When the temperature of the air is within a degree or two of the freezing point, and much snow falls, it frequently consists of large, irregular flakes, such as are common in Britain. Sometimes it exhibits small granular, or large rough, white concretions; at others it consists of white spiculæ, or flakes composed of coarse spiculæ, or rude, stellated crystals formed of visible grains. But in severe frosts, though the sky appears perfectly clear, lamellar flakes of snow of the most regular and beautiful forms are always seen floating in the air and sparkling in the sunbeams; and the snow which falls in general is of the most elegant texture and appearance."

Of the hidden causes which originate these beautiful productions, nothing whatever is known. Some have imagined that they are to be found in the forms of the primal atoms of water, which are assumed to be triangles or hexagons, and which, therefore, uniting by their similar sides or edges, must give rise to crystals of regular forms. Others find the solution in magnetic or electrical affinities, which are supposed to require the particles to unite by some law of polar attraction. But even if these theories were demonstrated, they would explain nothing. *Why* the particles must unite in these particular methods, or what is the nature of attraction itself, no man knows. It is sufficient to say, with the learned and devout navigator who has done most to make us acquainted with these beautiful objects, "Some of the general varieties in the figures of the crystals may be referred to the temperature of the air; but the particular and endless modifications of similar classes of crystals can only be referred to the will and pleasure of the great First Cause, whose works, even the most minute and evanescent, and in regions the most remote from human observation, are altogether admirable."

Snow is formed in the higher regions of our atmosphere. It is the wild, raging water of the ocean, the gentle rill of the mountains, the beautiful lake, and the vilest pond on earth, all taxed and made to contribute at the bidding of their Lord to this department of his treasure-house. They send up their tribute in the finest particles of moisture; the steady contribution coming up from all parts of the globe indiscriminately. No matter what king claims the fields and rivers and mountains to minister to his wants, our God makes them all fill his treasury. The vapor comes up like gold, in grains and nuggets. It must be cast into the King's furnace and formed into his coin, before he can use it. Now tell me how he makes snow out of vapor. You can answer it in one sentence,— by diminishing the heat. Easily said; but who can do it? A profound philosopher, in remarking on the magnificent glacial phenomenon of January, 1845, when for eight days there was one of the most wonderful displays of the effects of cold perhaps ever witnessed in our latitude, when the earth and every twig seemed covered with diamonds, says of it, "Job speaks of the balancing of the clouds as among the mysteries of ancient philosophy. But how much nicer the balancing and counter-balancing of the complicated agencies of the atmosphere, in order to bring out this glacial miracle in full perfection! What wisdom and power short of infinite could have brought it about?" It is equally appropriate to ask, What but infinite power could produce all the agencies and instruments needed in creating one flake of snow? The tiny creature says, as you examine it, —

"The hand that made me is divine!"

Kirk.

The First Snow.

LOVE to watch the first soft snow,
　As it slowly saileth down,
Purer and whiter than the pearls
　That grace a monarch's crown;
Though winter wears a freezing look
　And many a surly frown.

It lighteth like the feathery down
　Upon the naked trees,
And on the pale and withered flowers
　That swing in every breeze;
And they are clothed in such bright robes
　As summer never sees.

It bringeth pleasant memories —
　The falling, falling snow —
Of neighing steeds, and jingling bells,
　In the happy long ago,
When hopes were bright, and health was good,
　And the spirits were not low.

And it giveth many promises
　Of quiet joys in store,

Of bliss around the blazing hearth
   When daylight is no more, —
Such bliss as nowhere else hath lived
   Since the Eden days were o'er.

God bless the eye that views, with mine,
   The falling snow to-day;
May Truth her pure white mission spread
   Before its searching ray,
And lead, with dazzling garments, toward
   The "strait and narrow way."

Julia H. Scott.

Peboan.

NOW, o'er all the dreary North-land,
   Mighty Peboan, the Winter,
   Breathing on the lakes and rivers,
   Into stone had changed their waters;
From his hair he shook the snow-flakes
Till the plains were strewn with whiteness,
One uninterrupted level,
As if, stooping, the Creator
With his hand had smoothed them over.

Longfellow.

Unity in Diversity.

The Lord hath his way in the whirlwind and in the storm. — Nahum 1 : 3.

BEDIENCE to law is apparent in all the works of the Creator. However varied or complicated their structure, however intricate their motions, however multiform their aspects, there is an all-wise design pervading them, a clue running through all diversities, and reducing all to unity and harmony in the grand scheme of the universe. The Lord hath *his* way in them all; and that is the single line of righteousness and beneficence.

Amid the endless varieties of the snow crystals a singular law of unity is apparent. It is the *angle of sixty degrees,* or some multiple of it. This is one-sixth of the complete circle; hence the hexiform or six-sided configuration of its prisms and plates. It is curious to glance over the patterns which we exhibit, and trace the operations of this law. Let the congealing vapor assume what fantastic shapes it will, let it riot in the profusion of its beautiful efflorations, yet it can never escape the control of that central attraction which binds them all in one. Hence their name, *flakes*, *i. e.*, flocks; the fleecy crystals, though spreading abroad each in its utmost individual liberty, being still retained within one ownership and belonging to one fold.

Like this law of unity in nature is God's great law of love in his moral realm. It is the principle of order and harmony throughout his intelligent universe. God's own nature is love, and it reigns among all the shining ranks of heaven. And in the numberless worlds which fill immensity, and through the utmost variety of capacities and grades of beings, it needs but the fulfillment of this law to secure universal joy. Love is the one principle which binds all individuals and provinces of his rational kingdom to each other, and each to his throne.

One great law of crystallization controls the whole snow world. Every flake has a skeleton as distinct as the human skeleton, and yet the individual flake is as different from its neighbor as man is from his. The fundamental law of the snow is to crystallize in three, or some multiple of three. All its angles must be sixty, or one hundred and twenty. All its prisms and pyramids must be triangular or hexagonal; whether spicular, or pyramidal, or lamellar, it ever conforms to its own great law of order, and thus conveys delight to the eye, and most delight to him who, having pleasure in the works of God, searches them out.

Some men reproach the Protestant Church for its various sects. But let such men examine God's works. Unity in variety is the law of the snow. There is a Trinity in it. Every snow-flake imitates its Creator by being three in one. It has a stern basis of fundamental doctrine; and it would excommunicate any snow-flake that tried to stand on any other. But around that fundamental unity is the free play of individual peculiarities. All snow-flakes are alike essentially, while probably no two are identical in details.

Kirk.

The Snow Shower.

STILL falling, falling, falling fast,
These messengers have come at last,
All floating through the chilly air,
On softest pinions, white and fair,
Each like a dove with downy breast,
High fluttering o'er its icy nest.

So coming, coming, coming still
From heaven, what daily blessings fill
Life's chalice with full many a joy,
Which Time's cold hand can not destroy;
So pure, so holy at their birth,
They sweetly charm the ills of earth.

Upon my heart, when lone and still,
As freely may pure gifts distill,
Awakening strains of sweetest peace,
Whose melody shall never cease,
Till, far beyond the reach of time,
They swell heaven's harmony sublime.

Drown.

To a Snow-Flake.

THOU rain-drop, snow-crystalled, most fragile and fair!
   Borne from the far cloud-land and fashioned in air,

   What measures befit thee? Some song of the spheres
Should chant out thy praise, with the swift-rolling years.

Like a gem cast in setting from heaven's bright floor,
Thou art pure, and all perfect, like the One we adore.

Three stars make thy glory, — the mystical sign
Of the Three named in heaven, — One Spirit Divine!

I behold thee afar, lest a warm human touch,
And the breath of my singing, with praise over-much,

Should make thee to perish, thy grace disappear,
And thy soft beauty vanish, dissolved in a tear.

Like thee I have come from a realm far away,
And like thee I shall live but a brief mortal day.

Like thee I shall find neither haven nor rest,
And descend like thee, once, to the Earth's frozen breast;

But thou shalt move onward, completing God's will
Through the courses of Nature, thy round to fulfill;

A drop in the river — a flake mid the snows —
A gleam in the rainbow — the dew on the rose.

Then go, little snow-flake ! I prize thee not less,
That our paths must divide, and I onward must press;

For I must mount upward yet higher and higher,
Nevermore to descend with heart-baffled desire;

For the light of God's Being is kindled in mine,
And my soul in his presence for ever must shine,

Where the purest snow-crystal looks tarnished and dim,
By the white jasper walls; where the saints' choral hymn

Floats up day and night through the fair golden street,
God's praise on their lips, and their crowns at his feet,

Exchanging Earth's crosses, the frost and the blight,
The snows of the valley, and mists of the night,

The brightest earth-blossoms, all the world can afford,
For Sharon's red rose, and the smile of the Lord.

S. D. C.

PERFECTION.

Praise the Lord, Snow and Vapors, fulfilling his word. — Psalm 148 : 8.

STRIKING characteristic of the snow crystal is its perfection of form. Whatever be the type of its structure, that type is completed with the utmost regularity and nicety. Every angle is of the prescribed size,—not a degree more or less. The number of parts is uniform. You will never see a star with five rays, nor seven. With a precision which art would strive in vain to excel, the pattern is carried out in detail with the most exact symmetry, and in the most nicely-adjusted proportions.

It is so in *every* snow-crystal, unless broken or otherwise injured. God has no show specimens in his cabinets, elaborately finished, while the mass of them are left imperfect. There are no obscure corners, no back apartments, where the half-formed, ill-shapen, abortive portions of his work are gathered, out of sight. The tiniest speck that is lost in the countless multitudes that robe the earth is as perfect as if the skill of the Creator had been expended upon this alone. The flake that falls in the vast polar solitudes, where no eye of man will ever see it, or that plunges to instant death in the ocean, is wrought with as much care and fidelity as if it were to sparkle in a regal crown.

If there be apparent exceptions to this general statement, they are only apparent, and even these confirm the fact. In ordinary storms, large portions of the flakes are broken, sometimes reduced almost to shapeless dust. Often the flakes, coming in contact with each other, adhere, and constitute masses which are very irregular. Sometimes, however, this union gives rise to regular forms, as twelve-pointed stars, which are believed to be two hexagons, the one of them overlapping the other. (Page 11 fig. 22, 23.) A few cases have been observed where a ray or point of a star has become the germinating center of a twin or parasitic star, forming together a structure anomalous as a whole, though regular in each of its individual parts. (Page 99 fig. 2.)

This universal perfection of figure results from the constancy and uniformity of the laws which govern the process of crystallization. But it is not too much to go beyond these, and behold a Divine mind which loves beauty for its own sake, and delights to sow it broadcast throughout creation. Though there be no human eye to behold and to admire, they will not therefore be unbeheld. It is not true that

" Full many a flower is born to blush *unseen,*
And *waste* its sweetness on the desert air."

The universe is full of conscious intelligence, from him who is the " Father of lights," downward through endless ranks of being, and the hymn of admiring praise perpetually ascends to him for the perfection and glory of his works.

Obey God. His laws to the snow-flake are designed to make it beautiful and useful. So are his laws to you. He tells the flake to put on such a form and go to such a place, and it goes without murmuring or reluctance. Obey God, and you will put on the beauty of holiness and bless the world.   Kirk.

The Snow-Flake.

"NOW, if I fall, will it be my lot
To be cast in some low and lonely spot,
To melt, and to sink unseen and forgot?
 And there will my course be ended?"
'Twas thus a feathery snow-flake said,
As down through measureless space it strayed,
Or as, half by dalliance, half afraid,
 It seemed in mid-air suspended.

"Oh, no!" said the earth; "thou shalt not lie
Neglected and lone on my lap to die,
Thou pure and delicate child of the sky;
 For thou wilt be safe in my keeping.
But then, I must give thee a lovelier form;—
Thou wilt not be a part of the wintry storm,
But revive, when the sunbeams are yellow and warm,
 And the flowers from my bosom are peeping!

"And then thou shalt have thy choice, to be
Restored in the lily that decks the lea,
In the jessamine bloom, the anemone,
 Or aught of thy spotless whiteness;—
To melt, and be cast, in a glittering bead,
With the pearls that the night scatters over the mead,
In the cup where the bee and the fire-fly feed,
 Regaining thy dazzling brightness.

"I'll let thee awake from thy transient sleep,
When Viola's mild blue eye shall weep,
In a tremulous tear; or, a diamond, leap
 In a drop from the unlocked fountain;
Or, leaving the valley, the meadow, and heath,
The streamlet, the flowers, and all beneath,
Go up, and be wove in the silvery-wreath
 Encircling the brow of the mountain.

"Or, wouldst thou return to a home in the skies,
To shine in the Iris, I'll let thee arise,
And appear in the many and glorious dyes
 A pencil of sunbeams is blending;
But true, fair thing, as my name is Earth,
I'll give thee a new and a vernal birth,
When thou shalt recover thy primal worth,
 And never regret descending."

"Then I will drop!" said the trusting flake;
"But bear it in mind that the choice I make,
Is not in the flowers nor the dew to wake,
 Nor the mist that shall pass with morning;
For, things of thyself, they will die with thee;
But those that are lent from on high, like me,
Must rise, — and will live, from the dust set free,
 To the regions above returning.

"And, if true to thy word and just thou art,
Like the spirit that dwells in the holiest heart,
Unsullied by thee, thou wilt let me depart,
    And return to my native heaven.
For I would be placed in the beautiful bow,
From time to time in thy sight to glow,
So thou mayst remember the Flake of Snow,
    By the promise that God has given!"

                                     H. F. Gould.

Mabel's Wonder.

THERE must be flowers in heaven!"
    Little Mabel wondering cried,
    As she gazed through the frosted window,—
    "Ah yes, ah yes," I replied.

"And every single blossom
    Is white as white can be!"
"Perhaps;"—I carelessly answered,
    "When we get there, we shall see."

"And oh, they have so many!
    Why, every tree must be full!"
"Of course,—Spring lasts for ever
    In heaven," I said, so dull.

"Do the angels get tired of flowers?"
    Asked she, with a gentle sigh;
"For see, oh see, they are throwing
    Whole handfuls down from the sky!"

I sprang to the frosted window,
    To see what the child could mean :—
The ground was covered with snow-flakes,
    And the air was full between.

I kissed my innocent darling,
    And speedily set her right;
While I prayed that her heart might ever
    Be pure as the snow and as light.

H. E. B.

It Snows!

IT snows! it snows! from out the sky
    The feathered flakes, how fast they fly!
    Like little birds, that don't know why,
    They're on the chase from place to place,
While neither can another trace.
It snows! it snows! a merry play
Is o'er us on this heavy day.

Like dancers in an airy hall
That has not room to hold them all,
While some keep up, and others fall,
The atoms shift, then thick and swift
They drive along to form the drift
That, weaving up, so dazzling white,
Is like a rising wall of light.

But now the wind comes whistling loud,
To snatch and waft it as a cloud,
Or giant phantom in a shroud.
It spreads; it curls; it mounts and whirls;
At length a mighty wing unfurls;
And then, away! but where, none knows,
Or ever will. It snows! it snows!

To-morrow will the storm be done;
Then out will come the golden sun,
And we shall see upon the run
Before his beams, in sparkling streams,
What now a curtain o'er him seems.
And thus with life it ever goes —
'Tis shade and shine! — It snows! it snows!

                                        H. F. Gould.

PURITY.

URITY is one of the most striking characteristics of the new-fallen snow. "It is," says Sturm, "a result of the congregated reflections of light from the innumerable small faces of the crystals. The same effect is produced when ice is crushed to fragments. It is extremely light and thin, consequently full of pores, and these contain air; it is farther composed of parts more or less compact, and such a substance does not admit the sun's rays to pass, neither does it absorb them; on the contrary, it reflects them very powerfully, and this gives it the dazzling white appearance we see in it."

You shall look out upon a gray, frozen earth, and a gray, chilling sky. The trees stretch forth naked branches imploringly. The air pinches and pierces you; a homesick desolation clasps around your shivering, shrinking frame, and then God works a miracle. The windows of heaven are opened, and there comes forth a blessing. The gray sky unlocks her treasures, and softness and whiteness and warmth and beauty float gently down upon the evil and the good. Through all the long night, while you sleep, the work goes noiselessly on. Earth puts off her earthliness; and when the morning comes, she stands before you in the white robes of a saint. The sun hallows her with baptismal touch, and she is glorified. There is no longer on her pure brow any thing common or unclean. The Lord God hath wrapped her about with light as with a garment. His divine charity hath covered the multitude of her sins, and there is no scar or stain, no " mark of her shame," or " seal of her sorrow." The far-off hills swell their white purity against the pure blue of the heaven. The sheeted splendor of the fields sparkles back a thousand suns for one. The trees lose their nakedness and misery and desolation, and every slenderest twig is clothed upon with glory.

Wheat-fields, corn-fields, and meadow-lands are all alike wrapped by its dazzling mantle. Here and there some straggling weeds refuse to be hidden, and stand up in unsightly contrast with the pure white surface around them. The stone walls, entirely concealed, only look like a low ridge; but the snow can not contrive to cover up the rail-fences, but only heap up a bank by their side. The woods, with their bare trunks and intermingling branches, cast a shadow, notwithstanding the absence of leaves, and we are glad again to come into the warm sunshine. Evergreens do not brighten a winter landscape. They seem as if they were mourning in sympathy with their spoiled brethren of the forest; and look dusky and almost black, like somber sentinels along the road. The snow sparkles with its crystals. What purity! "Whiter than snow!" The longing of the soul for purity, the faith in the cleansing power that is able thus to purify, are breathed in the prayer, "Wash me, and I shall be whiter than snow!"

The deep, deep snow offers no temptations to wander in the fields, or step away from the beaten track. One well-defined road, from which the driver reluctantly turns aside on meeting another sleigh, has alone broken the crust on its surface, and determines its depth. Foot-passengers step out into the deep snow, and wait till the sleigh passes, but are glad at once to step back again. How well would it be if Christians thus dreaded to step aside from the narrow way that leadeth unto life, and were as ready to return to its secure footing, — the path beaten by blessed foot-prints!

Beecher.

What comes from heaven is pure; but the tendency is to soil it, and that which keeps nearest heaven most escapes the pollution of earth. At the foot of the Alps you find the roaring, muddy stream, the clay-stained snow. But on the summit of Mt. Blanc is a pure robe of celestial white, never stained, only sometimes covered with a roseate gauze to salute the setting sun.

Kirk.

The snow is very beautiful when it has first fallen. Many of our poets have had recourse to the snow-flake for some of their finest poetical images; nor do

I know a fitter emblem of innocence and purity than a falling flake, ere it receives the stain of earth. There are but few things with which we can compare snow for purity. The Psalmist says, "He giveth the snow like wool; he scattereth the hoar frost like ashes." "Wash me, and I shall be whiter than snow." Milton has made beautiful allusion to it in his hymn on the Nativity, where he says, —

> "It was the winter mild,
> While the heaven-born Child
> All meanly wrapped in the rude manger lies;
> Nature in her awe to Him
> Had doffed her gaudy trim,
> With her great Master so to sympathize.
> It was no season then for her
> To wanton with the sun, her lusty paramour.
>
> "Only with speeches fair
> She woos the gentle air
> To hide her guilty front with innocent snow;
> And on her naked shame,
> Pollute with sinful blame,
> The saintly vail of maiden white to throw,
> Confounded that her Maker's eyes
> Should look so near upon her foul deformities."
>
> T. Miller.

The Snow-Wreath.

OH, what is so pure, so soft, or so bright,
   As a wreath of the new-fallen snow!
It seems as if brushed from the garments of light,
   To fall on us mortals below.

But here is no home for thee, child of the sky;
   Thy purity here must decay,
Thy being be transient, thy beauty all die,
   Nor a trace of thy loveliness stay.

Go, go to the mountain-top, — there make thy nest;
   'Tis nearer thy own native home;
And live on its peak like a silvery crest,
   Where nothing to soil thee can come.

This emblem — how apt of a virtuous mind
   Made pure by the Spirit divine!
Like a snow-wreath 'tis marred in a world so unkind,
   Fit only in heaven to shine.

                 J. B. Waterbury.

The First Snow.

TO-DAY has been a pleasant day,
　　Despite the cold and snow:
A Sabbath stillness filled the air,
And pictures slumbered every where,
　　Around, above, below.

We woke at dawn and saw the trees
　　Before our windows white;
Their limbs were clad with snow-like bark,
Save that the under sides were dark,
　　Like bars against the light.

The fence was white around the house,
　　The lamp before the door;
The porch was glazed with pearlèd sleet,
Great drifts lay in the silent street,
　　The street was seen no more!

Long trenches had been roughly dug,
　　And giant foot-prints made;
But few were out; the streets were bare;
I saw but one pale wanderer there,
　　And he was like a shade!

I seemed to walk another world,
    Where all was still and blest;
The cloudless sky, the stainless snows, —
It was a vision of repose,
    A dream of heavenly rest, —

A dream the holy night completes;
    For now the moon hath come,
I stand in heaven with folded wings,
A free and happy soul that sings
    When all things else are dumb!

Richard Henry Stoddard.

THROUGH the hushed air the whitening shower
    descends,
    At first thin-wavering, till at last the flakes
    Fall broad and wide and fast, dimming the day
With a continual flow. The cherished fields
Put on their winter robe of purest white;
'Tis brightness all, save where the new snow melts
Along the mazy current. Low the woods
Bow their hoar head; and, ere the languid sun
Faint from the west emits its evening ray,
Earth's universal face, deep-hid and chill,
Is one wide dazzling waste, that buries wide
The works of man.

Thomson.

GRACE.

The Lord giveth Snow like wool. — Psalm 47 : 16.

COLD and dreary as winter is, it is not devoid of interest to the man of taste and Christian sentiment. Look at the delicate snow-flake. With what grace of motion has God endowed it! How childlike, gently, peacefully, confidingly the little creature comes down into our turbulent earth! It is not difficult to conceive that it comes as an attendant on some angel, whose movements it imitates. Kirk.

We have sat and watched the fall of snow until our head grew dizzy, for it is a bewitching sight to persons speculatively inclined. There is an aimless way of riding down, a simple, careless, thoughtless motion, that leads you to think that nothing can be more nonchalant than snow. And then it rests upon a leaf or alights upon the ground, with such a dainty step, so softly, so quietly, that you almost pity its virgin helplessness. If you reach out your hand to help it, your very touch destroys it. It dies in your palm, and departs as a tear.

Nowhere is snow so beautiful as when one sojourns in a good old-fashioned mansion in the country, bright and warm, full of home joy and quiet. You look out through large windows and see one of those flights of snow in a still, calm day, that make the air seem as if it were full of white millers or butterflies, fluttering down from heaven. There is something extremely beautiful in the motion of these large flakes of snow. They do not make haste, nor plump straight down with a dead fall, like a whistling raindrop. They seem to be at leisure; and descend with that quiet, wavering, sideway motion which birds sometimes use when about to alight. You think you are reading; and so you are, but it is not in the book that lies open before you. The silent, dreamy hour passes away, and you have not felt it pass. The trees are dressed with snow. The long arms of evergreens bend with its weight; the rails are doubled, and every post wears a starry crown. The well-sweep, the bucket, the well-curb are fleeced over. And still the silent, quivering air is full of trooping flakes, thousands following to take the place of all that fall. The ground is heaped, the paths are gone, the road is hidden, the fields are leveled, the eaves of buildings jut over, and, as the day moves on, the fences grow shorter, and gradually sink from sight. All night the heavens rain crystal flakes. Yet, that roof, on which the smallest rain pattered audible music, gives no sound. There is no echo in the stroke of snow, until it waxes to an avalanche and slips from the mountains. Then it fills the air like thunderbolts.

Beecher.

Falling snow is beautiful in a forest. It comes wavering down among the trees without a whisper, and takes to the ground without the sound of a foot-fall. Evergreen trees grow intense in contrasts of dark-green ruffled with radiant white. Bush and tree are powdered and banked up. Not the slight-est sound is made in all the work which fills the woods with winter soil many feet deep. When the

morning comes, then comes the sun also. The **storm**
has gone back to its northern nest to shed its feath-
ers there. The air is still, cold, bright. But what **a**
glory rests upon the too brilliant earth! **Are these**
the January leaves?— is this the winter efflorescence
of shrub and tree? You can scarcely look for the
exceeding brightness. Trees stand up against the
clear, gray sky, brown and white in contrast, **as if**
each trunk and bough and branch and twig **had**
been coated with ermine, or with **white moss.**
There is an exquisite airiness and lightness in the
masses of snow on trees and fences, when seen just
as the storm left them. The wind or sun soon dis-
enchants the magic scene.   **Ib.**

The Snow-Storm.

ANNOUNCED by all the trumpets of the sky
    Arrives the snow, and, driving o'er the fields,
    Seems no where to alight; the whited air
    Hides hills and woods, the river and the heaven,
And vails the farm-house at the garden's end.
The sled and traveler stopped, the courier's feet
Delayed, all friends shut out, the house-mates sit
Around the radiant fireplace, inclosed
In a tumultuous privacy of storm.
    Come see the north-wind's masonry.
Out of an unseen quarry, evermore
Furnished with tile, the fierce artificer
Carves his white bastions, with projected roof
Round every windward stake, or tree, or door,
Speeding, the myriad-handed, his wild work
So fanciful, so savage; naught cares he
For number or proportion. Mockingly
On coop or kennel he hangs Parian wreaths;
A swan-like form invests the hidden thorn;
Fills up the farmer's lane from wall to wall,
Mauger the farmer's sighs, and at the gate
A tapering turret overtops the work.
And when his hours are numbered, and the world
Is all his own, retiring as he were not,

Leaves, when the sun appears, astonished Art
To mimic, in those structures, stone by stone,
Built in an age, the mad wind's night-work,
The frolic architecture of the snow.

R. W. Emerson.

The Spirit of the Snow.

THE night brings forth the morn;
Of the cloud is lightning born;
From out the darkest earth the brightest roses grow;
Bright sparks from black flints fly;
And from out a leaden sky
Comes the silvery-footed Spirit of the Snow.

The wondering air grows mute,
As her pearly parachute
Cometh slowly down from heaven, softly floating to and fro;
And the earth emits no sound,
As lightly on the ground
Leaps the silvery-footed Spirit of the Snow.

At the contact of her tread,
The mountain's festal head
As with chaplets of white roses seems to glow;

And its furrowed cheeks grow white,
With a feeling of delight,
At the presence of the Spirit of the Snow.

As she wendeth to the vale,
The longing fields grow pale,
The tiny streams that vein them cease to flow;
And the river stays its tide,
With wonder and with pride,
To gaze upon the Spirit of the Snow.

But little does she deem
The love of field or stream;
She is frolicsome and lightsome· as the roe;
She is here and she is there;
On the earth or in the air,
Ever-changing, floats the Spirit of the Snow.

Now, a daring climber, she
Mounts the tallest forest tree,—
Out along the dizzy branches doth she go!
And her tassels, silver-white,
Down-swinging through the night,
Mark the pillow of the Spirit of the Snow.

Now she climbs the mighty mast,
Where the sailor-boy at last
Dreams of home, in his hammock down below;

There she watches in his stead,
Till the morning light shines red,
Then evanishes the Spirit of the Snow.

Or, crowning with white fire
The minster's topmost spire
With a glory such as sainted foreheads show,
She teaches fanes are given
Thus to lift the heart to heaven,
There to melt like the Spirit of the Snow.

Now above the loaded wain,
Now beneath the thundering train,
Doth she hear the sweet bells tinkle and the snorting en-
gine blow;
Now she flutters on the breeze,
Till the branches of the trees
Catch the tossed and tangled tresses of the Spirit of the
Snow.

Now an infant's balmy breath
Gives the Spirit seeming death,
When adown her pallid features fair, Decay's damp dew-
drops flow;
Now, again her strong assault
Can make an army halt,
And trench itself in terror 'gainst the Spirit of the Snow.

At times, with gentle power,
In visiting some bower,
She scarce will hide the holly's red, the blackness of the sloe;
But ah! her awful might,
When down some Alpine hight
The hapless hamlet sinks before the Spirit of the Snow.

On a feather she floats down
The turbid rivers brown,—
Down to meet the drifting navies of the winter-freighted foe;
Then swift o'er the azure walls
Of the awful waterfalls,
Where Niagara leaps roaring, glides the Spirit of the Snow.

With her flag of truce unfurled,
She makes peace o'er all the world,—
Makes bloody battle cease awhile, and war's unpitying woe;
Till its hollow womb within
The deep-mouthed culverin
Incloses, like a cradled child, the Spirit of the Snow.

In her spotless linen hood,
Like the other sisterhood,
She braves the open cloister where the psalm sounds sweet
                and low,
When some sister's bier doth pass
From the minster and the mass,
Soon to sink into the earth, like the Spirit of the Snow.

But at times so full of joy,
She will play with girl and boy,
Fly from out their tingling fingers, like white fire-balls on
the foe;
She will burst in feathery flakes;
And the ruin that she makes
Will but wake the crackling laughter of the Spirit of the
Snow.

Or, in furry mantle dressed,
She will fondle on her breast
The embryo buds awaiting the near Spring's mysterious throe
So fondly that the first
Of the blossoms that outburst
Will be called the beauteous daughter of the Spirit of the
Snow.

Ah! would that we were sure
Of hearts so warmly pure,
In all the winter weather that this lesser life must know;
That when shines the Sun of Love
From a warmer realm above,
In its life we may dissolve, like the Spirit of the Snow.

Dublin University Magazine.

BEAUTY.

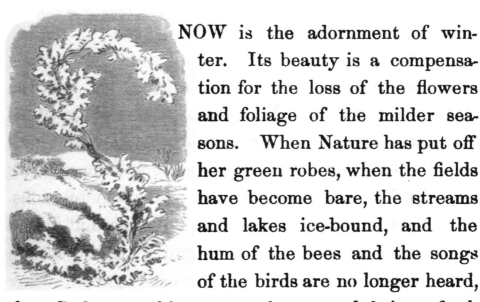

NOW is the adornment of winter. Its beauty is a compensation for the loss of the flowers and foliage of the milder seasons. When Nature has put off her green robes, when the fields have become bare, the streams and lakes ice-bound, and the hum of the bees and the songs of the birds are no longer heard, then God opens his treasure-house and brings forth jewels for the coronation of the year. He throws over the earth a robe of purest white, he festoons each shrub and tree with diamonds and pearls, and bids every beholder rejoice in these manifestations of his skill. For all the beauty of the earth is but the outward expression of the beauty which dwells eternal in the Divine Mind. Each six-leaved blossom of winter had its pattern in his thought before it was created; and all the diversities of its forms show the wealth of his resources even in the smallest things. So God has not left himself "without witness" for a single season. Each has its message from heaven, unfolding his glories, and bidding man behold and adore.

All the roofs are blanketed with snow; all the fences are bordered. Every gate-post is statuesque; every wood-pile is a marble quarry. Harshest outlines are softened. Instead of angles and ruggedness and squalor, there are billowy, fleecy undulations. Nothing so rough, so common, so ugly, but it has been transfigured into newness of life. Every where the earth has received beauty for ashes, the oil of joy for mourning, the garment of praise for the spirit of heaviness. Without sound of hammer or ax, without the grating of saw or the click of chisel, prose has been sculptured into poetry. The actual has put on the silver vail of the ideal. **Gail Hamilton.**

It is almost impossible to paint the glory of the northern winter forests. Every tree, laden with the purest snow, resembled a Gothic fountain of bronze, covered with frozen spray, through which only suggestive glimpses of its delicate tracery could be obtained. From every side we looked over thousands of such mimic fountains, shooting low or high, from their pavements of ivory and alabaster. It was an enchanted wilderness, — white, silent, gleaming, and filled with inexhaustible forms of beauty. To what shall I liken those glimpses under the boughs, into the depths of the forest, where the snow destroyed all perspective, and brought the remotest fairy nooks and coverts, too lovely and fragile to seem cold, into the glittering foreground? "Wonderful!" "Glorious!" I could only exclaim in breathless admiration.

Bayard Taylor.

The forests were indescribable in their silence, whiteness, and wonderful variety of snowy adornment. The weeping birches leaned over the road and formed white-fringed arches; the firs wore mantles of ermine, and muffs and tippets of the softest swan's down. Snow, wind, and frost had wrought the most

marvelous transformations.   Here were kneeling nuns
with their arms hanging listlessly by their sides, and
the white cowls falling over their faces.   There lay
a warrior's helmet; lace curtains, torn and ragged,
hung from the points of little Gothic spires.   Caverns,
lined with sparry incrustations, silver palm-leaves,
doors, loop-holes, arches, and cascades were thrown
together in fantastic confusion, and mingled with the
more decided forms of the larger trees, which were
trees but in form, so completely were they wrapped
in their dazzling disguise.   It was an enchanted land,
where you scarcely dared breathe, lest a breath might
break the spell.   ᴵᵇ.

The new snow had fallen on the mountains, and the
vast basin of the Monte Rosa chain lay before us,
clothed in flowing robes of the most pure and spotless
white ; while every little nook and ledge, and ine-
quality of rock on which the snow could rest, was
covered with the same virgin luster, so that it looked
as if the sides of the craggy mountains were flecked
and dashed with spray, and as if myriads of foaming
torrents were coursing down the precipices, streak-
ing their surface with their white tracks in every

direction. After we turned to the right and began the ascent, the light became stronger, and the outline sharper, and our view of the vast glacier basin more uninterrupted and clear. The valley of Macregnaga goes very far into the heart of the mountain, so that all the snowy part of Monte Rosa rises in one great mass directly above it. The sun came up, and for two or three minutes, not more, all the upper part of this vast region of snow was dyed of the deepest crimson, — not pink, as an evening view of the Alps often is; then, for much longer, it was of the most brilliant gold, — just the color of a new sovereign; and then, as the sun overtopped the lower mountains, and their shadows were no longer thrown upward, this gorgeous coloring gave place to a dazzling glare. Miles off, as we were, we could hardly look at the snowy basin without blinking.   Wills.

There is in us a want of taste to appreciate the exquisite beauty of the snow-flakes that we tread under foot. There is a narrow selfishness which does not even inquire what are the moral or æsthetic uses of the snow; but is contented or sad to see it come upon the earth, according as it affects our arrangements and wishes.   Our education has this radical defect, that it does not teach us to make the senses the instruments of our higher faculties; to study nature, to revere every thing that God makes; that it fails to form us to the highest exercises of which we are capable, and leaves us ignorant of some of the most interesting and important objects of knowledge: God, — his word, his works and ourselves.   **Kirk.**

Beauty.

THERE's beauty all around our paths,
　　If but our watchful eyes
　　Can trace it, mid familiar things,
　　　And through their lowly guise.
We may find it in the winter boughs,
　As they cross the cold blue sky,
While soft on icy pool and stream
　Their penciled shadows lie;
When we look upon their tracery,
　By the fairy frost-work bound,
Whence the flitting red-breast shakes a shower
　Of crystals to the ground.
　　　　　　　　　　Hemans.

The Beautiful Snow.

OH the snow! the beautiful snow!
Filling the sky and the earth below,—
Over the house-tops, over the street,
Over the heads of the people you meet.
 Dancing,
  Flirting,
   Skimming along,
Beautiful snow! it can do nothing wrong;
Flying to kiss a fair lady's cheek,
Clinging to lips in a frolicsome freak;
Beautiful snow from the heavens above,
Pure as an angel, and fickle as love!

Oh the snow! the beautiful snow!
How the flakes gather and laugh as they go!
Whirling about in its maddening fun;
It plays in its glee with every one.
 Chasing,
  Laughing,
   Hurrying by,
It lights up the face, and sparkles the eye;
And even the dogs, with a bark 'and a bound,
Snap at the crystals that eddy around:
The town is alive, and its heart in a glow,
To welcome the coming of beautiful snow.

How the wild crowd goes swaying along,
Hailing each other with humor and song!
How the gay sledges, like meteors flash by,
Bright for the moment, then lost to the eye!
   Ringing,
     Swinging,
       Dashing they go,
Over the crust of the beautiful snow,—
Snow so pure when it falls from the sky,
To be trampled in mud by the crowd rushing by,
To be trampled and tracked by thousands of feet,
Till it blends with the filth in the horrible street.

"Once I was pure as the snow—but I fell;
Fell like the snow-flake, from heaven—to hell;
Fell, to be trampled as filth of the street,
Fell to be scoffed, derided, and beat,
   Pleading,
     Cursing,
       Dreading to die,
Selling my soul to whomever would buy;
Dealing in shame for a morsel of bread,
Hating the living, and fearing the dead:
Merciful God! have I fallen so low?
And yet I was once like this beautiful snow!

"Once I was fair as this beautiful snow,
With an eye like its crystals, a heart like its glow;
Once I was loved for my innocent grace,
Flattered and sought for the charm of my face.
     Father,
         Mother,
             Sister, all,
God and myself I have lost by the fall.
The veriest wretch that goes shivering by
Will take a wide sweep lest I venture too nigh;
For of all that is on or about me, I know,
There is nothing that's pure but the beautiful snow.

"How strange it should be that this beautiful snow
Should fall on a sinner with no where to go!
How strange it would be, when the night comes again,
If the snow and the ice struck my desperate brain!
     Fainting,
         Freezing,
             Dying alone,
Too wicked for prayer, too weak for my moan
To be heard in the crash of the crazy town,
Gone wild in their joy at the snow's coming down,
To lie and to die in my terrible woe,
With a bed and a shroud in the beautiful snow!"

Helpless and foul as the trampled snow,
Sinner, despair not! Christ stoopeth low,
To rescue the soul that is lost in its sin,
And raise it to life and enjoyment again.
Groaning,
Bleeding,
Dying, for thee,
The Crucified hung on the accursed tree!—
His accents of mercy fall soft on thine ear.
"Is there mercy for me? Will he heed my weak prayer?
O God! in the stream that for sinners did flow,
Wash me, and I shall be whiter than snow!"

Selected.

Weakness.

The stream of brooks . . . . wherein the Snow is hid. — Job 6 : 15.

LIGHTNESS and weakness are symbolized by the snow. You can not draw near one of these delicate crystals without danger of destroying it. Your breath will melt it ; nay, even the radiation of warmth from your person will, ere you are aware, crumble down the whole fairy structure so elaborately wrought. It floats down, the sport of every breath of air. It can not ruffle the feather of a bird by its falling. It perishes if the sun looks at it. Yet God takes care of it, — numbers it among his treasures. It is not overlooked by him amid all its fellows. When it dies in a tear, God bottles that tear and keeps it still in his treasure-house. Fear not, ye of little faith; ye are of more value than mountains of snow-flakes. Does the Almighty create and delight in it, preserve and guide this little creature, and will he not take care of you, and delight to make you beautiful in holiness, and serviceable in his kingdom? Look up when it storms. The sun is on the other side. God guides the cloud, the wind, the rain, and the snow, and numbers the hairs of your head.

Kirk.

Do a little good at a time, and all the time. The Himalaya is ordered to put on a new robe. How is it to be done? Will a mighty vestment drop from heaven and encircle the mighty ranges of her peaks? No; millions of little maids of honor will come down, and each one contribute some little thread to weave the splendid robe. And by every one doing the little committed to it, the giant mountain stands robed in its celestial garment. You organize a Sunday school among neglected children, and go every Sunday, like a little snow-flake, to add present labor to past. Keep on; that is the way the Himalaya gets its robe.

No good is lost. Stop not to count your converts, to weigh the results of your labors, but keep on like the gentle snow, flake after flake, without noise or parade. Parent, teacher, preacher, patriot, work on!

Ib.

We see the instability of snow, and the rapidity with which it disappears when played upon by the sunbeams, or exposed to the effects of a humid, mild air, and frequent showers. Frequently the whole aspect of nature, in a few hours, assumes a new appearance, and scarcely a trace of snow is left behind. By these sudden changes we may justly be reminded of the inconstancy and vanity of all human affairs. Fleeting as the snow beneath the sunbeams are all the enjoyments and gratifications which do not arise from the influence of religion, the exercise of the mind, and the feelings of the heart; if we cultivate these, we shall be enabled to enjoy a portion of that felicity which endureth for ever, — the sure reward of virtue and a well-spent life.

Sturm.

Soon another silent force will come forth, and a noiseless battle will ensue, in which this now innumerable army of snow-flakes shall be itself vanquished.

A rain-drop is stronger than a snow-flake. One by one, the armed drops will dissolve the crystals and let forth the spirit imprisoned in them. Descending quickly into the earth, the drops shall search the roots, and give their breasts to myriad mouths. The bud shall open its eye, the leaf shall lift up its head, the grass shall wave its spear, and the forests hang out their banners! How significant is this silent, gradual, but irresistible power of rain and snow of moral truth in this world! "For, as the rain cometh down, and the snow from heaven, and returneth not thither, but watereth the earth, and maketh it bring forth and bud that it may give seed to the sower and bread to the eater; so shall my word be that goeth forth out of my mouth; it shall not return unto me void, but it shall accomplish that which I please, and it shall prosper in the thing whereto I sent it."

Beecher.

Nothing Lost.

WHERE is the snow?
        'Tis not long ago
It covered the earth with a vail of white;
We heard not its footsteps soft and light,
Yet there it was in the morning bright;
Now it hath vanished away from sight.
        Not a trace remains
        In fields or lanes.

Where is the frost?
It is gone and lost —
The forms of beauty last night it made;
With pictures rare were windows arrayed;
"Be silent!" it said; the brook obeyed;
Yet silence and pictures all did fade.
        At the smile of the sun,
        All was undone.

Where is the rain?
Pattering it came,
Dancing along with a merry sound,
A grassy bed in the fields it found;
Each drop came on the roof with a bound.
Where is the rain? It hath left the ground.
        What good hath it done —
        Gone away so soon?

Ever, ever,
Our best endeavor
Seemeth to fall like the melted snow.
We work out our thoughts wisely and slow;
The seed we sow, but it will not grow.
Our hopes, our resolves,—where do they go?
What doth remain?
Memory and pain.

Nothing is lost,—
No snow nor frost
That comes to enrich the earth again.
We thank them when the ripening grain
Is waving over the hill and plain,
And the pleasant rain springs from earth again.
All endeth in good,—
Water and food.

Never despair;
Disappointment bear;
Though hope seemeth vain, be patient still;
Thy good intent God doth fulfill;
Thy hand is weak; his powerful will
Is finishing thy life-work still.
The good endeavor
Is lost—ah! never.

Selected.

Snow-Flakes.

SEE the feathery snow-flakes
  Falling from the sky!
Myriads; yet so softly,
  There's no sound or sigh.
Mother Earth's brown raiment
  Doffs she for her white;
For a fairy's wand, transforming,
  Hides it out of sight.

Fairy little snow-flakes,
  Dancing as ye fall,
Resting on the rough old rock
  By the garden wall,
There's no spot so dreary,
  Naught so black and cold,
But your mantle may o'erspread
  With its falling fold.

Starry little snow-flakes,
  Blossoms of the sky,
Blooming when earth's daisies
  Fast asleep all lie,
Are ye beauty's raiment,
  On her bridal morn?
Or the floating garments
  By the frost fays worn?

Tell me, mystic snow-flakes,
   Is your home so far
You can hear the singing
   Of the morning star? —
Hear the grand, sweet chorus,
   As the spheres move on,
In their slow, majestic march
   Round the great white throne?

What are ye, fair snow-flakes,
   To the King of kings?
Unto Him who walketh
   On the wind's swift wings,
Maketh clouds his chariot,
   Light no man can see
Weareth for a garment, —
   Snow-flakes, what are ye?

Of his spotless purity
   But a shadow dim
And the silence of our coming
   Speaketh, too, of Him.
Mortals, stay your tear-drops!
   One day you shall know
What it is to be, like Him,
   "Whiter than the snow."

H. Maude H.

The Snow-Shower.

STAND here by my side, and turn, I pray,
  On the lake below thy gentle eyes;
The clouds hang over it heavy and gray,
  And dark and silent the water lies;
And out of that frozen mist the snow
In wavering flakes begins to flow;
      Flake after flake,
They sink in the dark and silent lake.

See how in a living swarm they come
  From the chambers beyond that misty vail;
Some hover awhile in the air, and some
  Rush from the sky like summer hail.
All, dropping swiftly or settling slow,
Meet and are still in the depth below;
      Flake after flake,
Dissolved in the dark and silent lake.

Here, delicate snow-stars, out of the cloud,
  Come floating downward in airy play,
Like spangles dropped from the glistening crowd
  That whitens by night the milky way;
There broader and burlier masses fall:
The sullen water buries them all,
      Flake after flake,
All drowned in the dark and silent lake.

And some, as on tender wings they glide
   From their chilly birth-cloud dim and gray,
Are joined in their fall, and, side by side,
   Come clinging along their unsteady way;
As friend with friend, or husband with wife,
Makes hand in hand the passage of life:
         Each mated flake
Soon sinks in the dark and silent lake.

Lo! while we are gazing, in swifter haste
   Stream down the snows till the air is white,
As, myriads by myriads madly chased,
   They fling themselves from their shadowy hight;
The fair frail creatures of middle sky,
What speed they make with their grave so nigh;
         Flake after flake
To lie in the dark and silent lake!

I see in thy gentle eyes a tear;
   They turn to me in sorrowful thought;
Thou thinkest of friends, the good and the dear,
   Who were for a time and now are not;
Like these fair children of cloud and frost,
That glisten a moment, and then are lost,
         Flake after flake, —
All lost in the deep and silent lake.

Yet look again, for the clouds divide;
  A gleam of blue on the mountain lies,
And far away on the mountain-side
  A sunbeam falls from the opening skies;
But the hurrying host, that flew between
The cloud and the water, no more is seen;
    Flake after flake
At rest in the dark and silent lake.

William Cullen Bryant.

Questions and Answers.

PRETTY little snow-flake,
  Floating softly by,
Bringest thou a message
  From the fleecy sky?

Yes, ah, yes, a lesson
  Beautiful as true;
Silent be, but busy,
  When you've work to do.
Avalanche and snow-drift
  Grow from single flakes;
Every crystal helping,
  Yet no noise it makes.

Glittering little snow-flakes,
  White as white can be,
How can I be spotless,
  Chaste, and pure, like thee?

All that comes from heaven
  Perfect is, like God;
But, alas! the sinner
  Earthly ways has trod;
Yet, to God returning,
  Thence anew to grow,
Sins, though they be scarlet,
  Shall be white as snow.

Loving little snow-flake,
  Tender is thy tread,
Weaving o'er the flowers
  Dainty coverlet.

Loving work is ever
  Best, when gently done;
All that's hard and selfish,
  Rough and cruel, shun.
Do each little duty
  With a smiling face,
Gathering all around you
  In love's warm embrace.

H. E. B.

POWER

Persecute them with thy tempest and make them afraid with thy storm.

Psalm 83 : 15.

F any one should ask what is the most harmless and innocent thing on earth, he might be answered, a snow-flake. And yet, in its own way of exerting itself, it stands among the foremost powers on earth. When it fills the air, the sun can not shine, the eye becomes powerless; neither hunter, nor pilot, guide nor watchman, is any better than a blind man. The eagle and the mole are on a level of vision. All the kings of the earth could not send forth an edict to mankind, saying, "Let labor cease." But this white-plumed light infantry clears out the fields, drives men home from the highway, and puts half a continent under ban. It is a despiser of old landmarks, and very quietly unites all properties, covering up fences, hiding paths and roads, and doing in one day a work which the engineers and laborers of the whole earth could not do in years !

But let the wind arise, and how is this peaceful seeming of snow-flakes changed ! In an instant the air raves. There is fury and spite in the atmosphere. It pelts you and searches you out in every fold and seam of your garments. It comes without search-warrant into each crack and crevice of your house. It pours over the hills, and lurks down in valleys, or roads, or cuts, until in a night it has entrenched itself formidably against the most expert human strength; for now, lying in drifts huge and wide, it bids defiance to engine and engineer. Before it this wonderful engine is as tame as a wounded bird; all its spirit is gone. No blow is struck. The snow puts forth no power. It simply lies still. That is enough. The laboring engine groans and pushes ; backs out and plunges in again; retreats and rushes again. It becomes entangled. The snow is every where. It is before it and behind it. It penetrates the whole engine, is sucked up in the draft, whirls in sheets into the engine-room; torments the cumbered wheels, clogs the joints, and, packing down under the drivers, it fairly lifts the ponderous engine from its feet, and strands it across the track! Well done, snow! That was a notable victory!

Beecher.

Look at the gentle flake coming down so silently, and then turn to contemplate its prodigious effects. Parent of a thousand of the streams and rivers that water and fertilize our globe, the snow-flake is equally the parent of the thundering avalanche that at St. Bernard overwhelms the unhappy traveler before he reaches the hospitable convent. In the afternoon, you find yourself suddenly caught in a storm. What is it that eclipses the sun, hours before his setting, that hides every landmark from the sight of the anxious guide, that turns day into sudden night? It is the snow-flake; for in the little thing is the hiding of God's power.

And is there not wealth as well as power in the enormous quantity of this one form of treasure lavished on the earth in one year? In one night you have found the earth covered with a carpet two feet in thickness. But if it requires millions of flakes for one cubic foot, what must it require to cover half the breadth of a continent on a meridian line of one thousand miles? And if that is repeated several months in each year, the mind staggers in the attempt at computation. "He giveth his snow like wool!" He scatters his pearls and diamonds by innumerable millions upon the earth. What prodigality of bounty our King displays!

The sovereign God gives the snow. It comes when he pleases, and falls where it pleases him to have it, on your house and your land; and you have no title that can prevent or bar his right. Napoleon may be the dread of kings, the mightiest monarch and warrior of the earth. He may be stronger than Russia, and may penetrate as far as Moscow. But Jehovah will there put a bridle in his mouth and a hook in his nostrils, and turn him backward, baffled, broken, disgraced. And he wanted for an instrument to accomplish his purposes the army of snow-flakes. He laid the deep covering of snow upon the earth; and the mighty army found themselves conquered by this little, gentle, silent instrument of God's power. God could have sent one warm storm of rain, and set the French army free. But he did not. He ruleth in the armies of heaven and doeth his pleasure among the inhabitants of the earth.   **Kirk.**

Scene in a Vermont Winter.

'TIS a fearful night in the winter-time,
      As cold as it ever can be;
The roar of the blast is heard like the chime
      Of the waves on an angry sea.
The moon is full; but her silver light
The storm dashes out with its wings to-night;
And over the sky from south to north
Not a star is seen, as the wind comes forth
      In the strength of a mighty glee.

All day had the snow come down, — all day, —
      As it never came down before;
And over the hills, at sunset, lay
      Some two or three feet or more;
The fence was lost, and the wall of stone;
The windows blocked, and the well-curbs gone;
The haystack had grown to a mountain lift,
And the wood-pile looked like a monster drift,
      As it lay by the farmer's door.

The night sets in on a world of snow,
      While the air grows sharp and chill,
And the warning roar of the fearful blow
      Is heard on the distant hill:

And the Norther, see! on the mountain-peak,
In his breath how the old trees writhe and shriek!
He drives from his nostrils the blinding snow;
He shouts on the plains, Ho-ho! ho-ho!
    And growls with a savage trill.

Such a night as this to be found abroad,
    In the drifts and the freezing air!
Sits a shivering dog, in the field, by the road,
    With the snow in his shaggy hair.
He shuts his eyes to the wind and growls;
He lifts his head and moans and howls;
Then crouching low, from the cutting sleet,
His nose is pressed on his quivering feet, —
    Pray what does the dog do there?

A farmer came from the village plain,
    But he lost the traveled way;
And for hours he trod with might and main
    A path for his horse and sleigh,
But colder still the cold winds blew,
And deeper still the deep drifts grew,
And his mare, a beautiful Morgan brown,
At last in her struggles floundered down,
    Where a log in a hollow lay.

In vain, with a neigh and a frenzied snort,
    She plunged in the drifting snow,
While her master urged, till his breath grew short,
    With a word and a gentle blow.
But the snow was deep and the tugs were tight;
His hands were numb and had lost their might;
So he wallowed back to his half-filled sleigh,
And strove to shelter himself till day,
    With his coat and the buffalo.

He has given the last faint jerk of the rein,
    To rouse up his dying steed;
And the poor dog howls to the blast in vain
    For help in his master's need.
For a while he strives, with a wistful cry,
To catch a glance from his drowsy eye,
And wags his tail if the rude winds flap
The skirt of the buffalo over his lap,
    And whines when he takes no heed.

The wind goes down and the storm is o'er —
    'Tis the hour of midnight, past,
The old trees writhe and bend no more
    In the whirl of the rushing blast.

The silent moon, with her peaceful light,
Looks down on the hills with snow all white,
And the giant shadow of Camel's Hump,
The blasted pine and the ghostly stump,
    Afar on the plain are cast.

But cold and dead, by the hidden log,
    Are they who came from the town, —
The man in his sleigh and his faithful dog
    And his beautiful Morgan brown
In the white snow desert, far and grand,
With his cap on his head and the reins in his hand —
The dog with his nose on his master's feet,
And the mare half seen through the crusted sleet
    Where she lay when she floundered down.

              Charles Gamage Eastman.

The Pass of the Sierra.

ALL night above their rocky bed
    They saw the stars march slow;
The wild Sierra overhead,
    The desert's death below.

The Indian from his lodge of bark,
  The gray bear from his den,
Beyond their camp-fire's wall of dark,
  Glared on the mountain men.

Still upward turned, with anxious strain,
  Their leader's sleepless eye,
Where splinters of the mountain-chain
  Stood black against the sky.

The night waned slow; at last, a glow,
  A gleam of sudden fire,
Shot up behind the walls of snow,
  And tipped each icy spire.

"Up, men!" he cried. "Yon rocky cone,
  To-day, please God, we'll pass,
And look from Winter's frozen throne
  On Summer's flowers and grass!"

They set their faces to the blast,
  They trod the eternal snow,
And, faint, worn, bleeding, hailed at last
  The promised land below.

Behind, they saw the snow-cloud tossed
  By many an icy horn;
Before, warm valleys, wood-embossed,
  And green with vines and corn.

They left the Winter at their backs,
  To flap his baffled wing,
And downward, with the cataracts,
  Leaped to the lap of Spring.

Strong leader of that mountain band!
  Another task remains:
To break from Slavery's desert land
  A path to Freedom's plains!

The winds are wild, the way is drear,
  Yet, flashing through the night,
Lo! icy ridge and rocky spear
  Blaze out in morning light!

Rise up, Frémont! and go before;
  The hour must have its man;
Put on the hunting-shirt once more,
  And lead in Freedom's van!

Whittier.

2

Gladness.

OW delightful is the face of nature when the morning light first dawns upon a country embosomed in snow! The thick mist which obscured the earth and concealed every object from our view at once vanishes. How beautiful are the tops of the trees, hoary with frost! The hills and valleys, reflecting the sunbeams, assume various tints; all nature is animated by the genial influence of the brightness, and, robed in white, delights the traveler with her novel and delicate appearance. How beautiful to see the white hills, the forests, and the groves all sparkling! What a delightful combination these objects present! Observe the brilliancy of the hedges! See the lofty trees bending beneath their dazzling burden! The surface of the earth appears one vast plain mantled in white and splendid array.   Sturm.

Already snow-birds are fluttering for a foothold, and showering down the frosty dust from the twigs. The hens and their uplifted lords are beginning to wade with dainty steps through the chilly wool. Boys are aglee with sleds; men are out with shovels, and dames with brooms. Bells begin to ring along the highway, and heavy oxen with craunching sleds are wending toward the woods for the winter's supply of fuel. The school-house is open, and a roasting fire rages in the box-stove. Little boys are crying with chilblains, and little girls are comforting them with the assurance that it will "stop aching pretty soon," and the boys seem unwilling to stop crying until then. Big boys are shaking their coats, and stamping off the snow, which peels easily from sleek, black-balled boots, or shoes burnished with tallow. Out of doors, the snowballs are flying, and every body laughs but the one that's hit. Down go the wrestlers. The big ones "rub" the little ones; the little ones in turn "rub" the smaller ones. The passers-by are pelted; and many a lazy horse has motives of speed applied to his lank sides. Even the schoolmaster is but mortal, and must take his lot; for many an "accidental" snowball plumps into his breast and upon his back before the rogues will believe that it *is* the schoolmaster.

But days go by. The snow drifts, — fences are banked up ten feet high. Hills are broken into a "coast" for boys' sleds. They slide and pull up again, and toil on in their slippery pleasure. They tumble over and turn over; they break down, or smash up; they run into each other, or run races, in all the moods and experiences of rugged frolic. Then comes the digging of chambers in the deep drifts, room upon room, the water dashed on overnight freezing the snow-walls into solid ice. Forts are also built, and huge balls of snow rolled up, till the little hands can roll the mass no longer.   Beecher.

For two days it had been storming. The air was murky and cross. The snow was descending, not peacefully and dreamily, but whirled and made wild by fierce winds. The forests were laden with snow, and their interior looked murky and dreadful as a witch's den. Through such scenes I began my ride upon the plow-shoving engine. The engineers and firemen were coated with snow from head to foot, and looked like millers who had not brushed their coats for ten years. The floor on which we stood was ice and snow half melted. The wood was coated with snow. The locomotive was frosted all over with snow, — wheels, connecting-rods, axles, and every thing but the boiler and smoke-stack. The side and front windows were glazed with crusts of ice, and only through one little spot in the window over the boiler could I peer out to get a sight of the plow. The track was indistinguishable. There was nothing to the eye to guide the engine in one way more than another. It seemed as if we were going across fields and plunging through forests at random. And this gave no mean excitement to the scene, when two ponderous engines were apparently driving us in such an outlandish excursion. But their feet were sure, and unerringly felt their way along the iron road, so that we were held in our courses.

Nothing can exceed the beauty of snow in its own organization, in the gracefulness with which it falls, in the molding of its drift-lines, and in the curves

which it makes when streaming off on either side
from the plow. It was never long the same. If the
snow was thin and light, the plow seemed to play
tenderly with it, like an artist doing curious things
for sport, throwing it in exquisite curves that rose
and fell, quivered and trembled as they ran. Then
suddenly striking a rift that had piled across the
track, the snow sprang out, as if driven by an explo-
sion, twenty and thirty feet, in jets and bolts; or like
long-stemmed sheaves of snow, — outspread, fan-like.
Instantly, when the drift was passed, the snow
seemed by an instinct of its own to retract, and
played again in exquisite curves, that rose and fell
about our prow. " Now you'll get it," says the
engineer, " in that deep cut." We only saw the first
dash, as if the plow had struck the banks of snow
before it could put on its graces, and shot it dis-
tracted and headlong up and down on either side,
like spray or flying ashes. It was but a second.
For the fine snow rose up around the engine, and
covered it in like a mist, and, sucking round, poured
in upon us in sheets and clouds, mingled with the
vapor of steam, and the smoke which, from impeded
draft, poured out, filled the engine-room and darkened
it so that we could not see each other a foot distant,

except as very filmy specters glowering at each other. Our engineers had on buffalo coats whose natural hirsuteness was made more shaggy by tags of snow melted into icicles. To see such substantial forms changing back and forth into a spectral lightness, as if they went back and forth between body and spirit, was not a little exciting to the imagination.

When we struck deep bodies of snow, the engine plowed through them laboriously, quivering and groaning with the load, but shot forth again, nimble as a bird, the moment the snow grew light and thin.

Nothing seemed wilder than to be in one of these whirling storms of smoke, vapor, and snow, you on one ponderous monster, and another roaring close behind, both engines like fiery dragons harnessed and fastened together, and looming up when the snow and mists opened a little, black and terrible. It seemed as if you were in a battle. There was such energetic action, such irresistible power, such darkness and light alternating, and such fitful half-lights, which are more exciting to the imagination than light or darkness. Thus, whirled on in the bosom of a storm, you sped across the open fields, full of wild-driving snow; you ran up to the opening of the

black pine and hemlock woods, and plunged into
their somber mouth as if into a cave of darkness,
and wrestled your way along through their dreary
recesses, emerging to the cleared field again, with
whistles screaming and answering each other back
and forth from engine to engine.   **Ib.**

It is not only that the snow makes fair what was
good before, but it is a messenger of love from heav-
en, bearing glad tidings of great joy. Hope for the
future comes down to the earth in every tiny snow-
flake. Underneath, as they span the hill-side, and lie
lightly piled in the valleys, the earth-spirits and
fairies are ceaselessly working out their multifold
plans. The grasses hold high carnival safe under
their crystal roof. The roses and lilies keep holiday.
The snow-drops and hyacinths, and the pink-lipped
May-flower, wait as they that watch for the morning.
The life that stirs beneath thrills to the life that stirs
above. The spring sun will mount higher and high-
er in the heavens; the sweet snow will sink down
into the arms of the violets, and, at the word of the
Lord, the Earth shall come up once more as a bride
adorned for her husband.   **Gail Hamilton.**

The Time of Snow.

BRAVE Winter and I shall ever agree,
Though a stern and frowning gaffer is he.
I like to hear him, with hail and rain,
Come tapping against the window-pane;
I like to see him come marching forth,
Begirt with the icicle-gems of the north;
But I like him best, when he comes bedight
In his velvet robes of stainless white.

A cheer for the snow, — the drifting snow,
Smoother and purer than Beauty's brow!
The creature of thought scarce likes to tread
On the delicate carpet daintily spread.
With feathery wreaths the forest is bound,
And the hills are with glittering diamonds crowned.
'Tis the fairest scene we can have below,
Sing a welcome, then, to the drifting snow.

The urchins gaze with eloquent eye,
To see the flakes go dancing by.
In the blinding storm how happy are they,
To welcome the first deep, snowy day!
Shouting and pelting, what bliss to fall,
Half-smothered, beneath the well-aimed ball!

Men of fourscore, did ye ever know
Such sport as ye had in the drifting snow?
Ye rejoice in it still, and love to see
The ermine mantle on tower and tree.
'Tis the fairest scene we can have below,
Hurrah! then hurrah! for the drifting snow!

Eliza Cook.

A Winter Sketch.

'TIS winter, yet there is no sound
    Along the air,
  Of winds upon their battle-ground;
    But gently there
The snow is falling,—all around,
    How fair,—how fair!

The jocund fields would masquerade;
    Fantastic scene!
Tree, shrub and lawn and lonely glade
    Have cast their green,
And joined the revel, all arrayed
    So white and clean.

E'en the old posts, that hold the bars,
   And the old gate,
Forgetful of their wintry wars
   And age sedate,
High-capped and plumed, like white hussars,
   Stand there in state.

The drifts are hanging by the sill,
   The eaves, the door;
The hay-stack has become a hill;
   All covered o'er
The wagon loaded for the mill
   The eve before.

Maria brings the water-pail,
   But where's the well?
Like magic of a fairy tale,
   Most strange to tell,
All vanished, curb and crank and rail!
   How deep it fell!

The wood-pile, too, is playing hide;
   The axe, the log,
The kennel of that friend so tried, —
   The old watch-dog, —
The grindstone standing by its side,
   All now *incog.*

The bustling cock looks out aghast
   From his high shed;
No spot to scratch him a repast;—
   Up curves his head,
Starts the dull hamlet with a blast,
   And back to bed.

Old drowsy dobbin, at the call,
   Amazed, awakes;
Out from the window of his stall
   A view he takes;
While thick and faster seem to fall
   The silent flakes.

The barn-yard gentry, musing, chime
   Their morning moan;
Like Memnon's music of old time,—
   That voice of stone!
So warbled they, and so sublime
   Their solemn tone.

Good Ruth has called the younker-folk
   To dress below;
Full welcome was the word she spoke;
   Down, down they go,
The cottage quietude is broke,—
   The snow!—the snow!

Now rises from around the fire
  A pleasant strain;
Ye giddy sons of mirth, retire!
  And ye profane!
A hymn to the Eternal Sire
  Goes up again.

The patriarchal Book divine,
  Upon the knee,
Opes where the gems of Judah shine,
  (Sweet minstrelsy!)
How soars each heart with each fair line,
  O God, to Thee!

Around the altar low they bend,
  Devout in prayer;
As snows upon the roof descend,
  So angels there
Come down that household to defend
  With gentle care.

While mounts the eddying smoke amain
  From many a hearth,
And all the landscape rings again
  With rustic mirth,
So gladsome seems to every swain
  The snowy earth.

Hoyt.

Gloom.

Hast thou seen the treasures of the hail, which I have reserved against the time of trouble? — Job 38 : 22.

OT in his splendors, only, nor his beneficence, does God manifest himself to men. "The Lord Most High is *terrible*." His holiness is for ever arrayed in frowns and rebuke against wrong. It is pleasant to dwell on his love, to speak of him as the Father of all his creatures, full of pity and condescension for the most erring. The heart that responds to his with reciprocal affection, penitent for sins committed, trustful in his promises of pardon through a Redeemer, and constrained by filial devotion to grateful service and worship, may rest in the sweet contemplation of his goodness. But let it not be forgotten at the same time that he is holy as well as good, "merciful and gracious, long-suffering, and abundant in goodness and truth, forgiving iniquity and transgression and sin, and that he will by no means clear the guilty."

It is fitting that this part, also, of the Divine character should be illustrated in his works. Therefore, he hath appointed the earthquake, the lightning, and the tempest, to be, with the sunshine and the gentle breezes, representatives of himself. Even the snow, so soft and beautiful, he makes a messenger of gloom. The dark, fierce winter storm sweeps over the earth as the very spirit of desolation. The tiny flakes, charged with the mission which he gives them, fly forth in numbers infinite to buffet, to bewilder, to overwhelm whatever is exposed to them. Who can resist these "treasures" of the storm when let loose in their strength? "Who can stand before his cold?"

As thus the snows arise, and foul and fierce
All Winter drives along the darkened air,
In his own loose, revolving fields the swain
Disastered stands; —— and wanders on
From hill to dale, still more and more astray,

Impatient, flouncing through the drifted heaps,
Stung with the thoughts of home, — the thoughts of
    home
Rush on his nerves, and call their vigor forth
In many a vain attempt.
                  How sinks his soul!
What black despair, what horror, fills his mind!
When for the dusky spot which fancy feigned
His tufted cottage rising through the snow,
He meets the roughness of the middle waste,
Far from the track and blest abode of man!
While round him Night resistless closes fast,
And every tempest, howling o'er his head,
Renders the savage wilderness more wild.
Nor wife nor children more shall he behold,
Nor friends nor sacred home. On every nerve
The deadly Winter seizes; shuts up sense;
And .o'er his inmost vitals creeping cold,
Lays him along the snows a stiffened corse,
Stretched out, and bleaching in the northern blast.

Thomson.

Winter is a fitting image of decay and death; and the cold, white, winding-sheet that shrouds the blighted flowers, of the robe that is spread over the still form of our heart's crushed and faded blossoms. True, the flowers shall spring again, and our treasures will be restored to us in the land of

eternal summer. Nevertheless, the dreary hour of
separation is not joyous, but grievous. Our hopes
are withered, our hearts are chilled; disappoint-
ment, absence, present loss, distress and torture **us.**
All is gloom, desolation, and anguish. The **sun**
may shine, but his beams are cold and **glassy.** No
pleasures spring about our path. Our life is **buried**
with our darlings in the icy bosom of **nature.** O
Winter, bleak, dismal, wasting, inexorable **Winter!**
Hasten thy footsteps, and bring us to **the bright**
expectant Spring!

Winter.

OH the long and dreary Winter!
   Oh the cold and cruel Winter!
   Ever thicker, thicker, thicker
   Froze the ice on lake and river;
Ever deeper, deeper, deeper
Fell the snow o'er all the landscape,—
Fell the covering snow, and drifted
Through the forest, round the village.
  Hardly from his buried wigwam
Could the hunter force a passage;

With his mittens and his snow-shoes
Vainly walked he through the forest,
Sought for bird or beast and found none, —
Saw no track of deer or rabbit;
In the snow beheld no footprints;
In the ghastly gleaming forest
Fell, and could not rise from weakness,
Perished there from cold and hunger.

Oh the famine and the fever!
Oh the wasting of the famine!
Oh the blasting of the fever!
Oh the wailing of the children!
Oh the anguish of the women!

All the earth was sick and famished;
Hungry was the air around them,
Hungry was the sky above them,
And the hungry stars in heaven,
Like the eyes of wolves glared at them!

Then they buried Minnehaha;
In the snow a grave they made her,
In the forest deep and darksome,
Underneath the moaning hemlocks;
Clothed her in her richest garments, —
Wrapped her in her robes of ermine,
Covered her with snow, like ermine;
Thus they buried Minnehaha.

Longfellow.

The Path through the Snow.

BARE and sunshiny, bright and bleak,
Rounded cold as a dead maid's cheek,
Folded white as a sinner's shroud,
Or wandering angel's robes of cloud, —
    Well I know, well I know,
Over the fields the path through the snow.

Narrow and rough it lies between
Wastes where the wind sweeps, biting keen;
Every step of the slippery road
Tracks where some weary foot has trod;
    Who will go, who will go,
After the rest, on the path through the snow?

They who would tread it must walk alone,
Silent and steadfast, one by one;
Dearest to dearest can only say,
"My heart! I'll follow thee all the way,
    As we go, as we go,
Each after each on this path through the snow."

It may be under that western haze
Lurks the omen of brighter days;

That each sentinel tree is quivering
Deep at its core with the sap of spring;
    And while we go, while we go,
Green grass-blades pierce through the glittering snow.

It may be the unknown path will tend
Never to any earthly end, —
Die with the dying day obscure,
And never lead to a human door;
    That none may know who did go
Patiently once on this path through the snow.

No matter, no matter! the path shines plain;
These pure snow-crystals will deaden pain;
Above, like stars in the deep blue dark,
Eyes that love us look down and mark.
    Let us go, let us go,
Whither Heaven leads in the path through the snow!

Miss Muloch.

December Snow.

FALL thickly on the rose-bush,
    O faintly falling snow!
For she is gone who trained its branch,
    And wooed its bud to blow.

Cover the well-known pathway,
  O damp December snow!
Her step no longer lingers there,
  When stars begin to glow.

Melt in the rapid river,
  O cold and cheerless snow!
She sees no more its sudden wave,
  Nor hears its foaming flow.

Chill every song-bird's music,
  O silent, sullen snow!
I can not hear her loving voice,
  That lulled me long ago.

Sleep on the earth's broad bosom,
  O weary winter snow!
Its fragrant flowers and blithesome birds
  Should with its loved one go.

                              W. B. Glazier.

Beneficence

The Rain cometh down and the Snow from heaven, and watereth the earth, and maketh it bring forth and bud, that it may give seed to the sower and bread to the eater. — Isaiah 55 : 10

HE great design of the snow is benevolent. It is appointed to water the earth, but not like the rain. That comes down and produces its effects and passes away, and is absent just when the heat is at its hight, and evaporation most rapid. But the snow comes down in the winter, and lies upon the high mountain ranges all through the hottest weather, gradually supplying the streams and rivers on which human life depends. The great Father of Waters, our grand Mississippi, is a child of the snow ; and often his waters swell when other streams are drying up. The very heat which is drinking up their waters is melting the snows on the Rocky Mountains, and replenishing his wasted bulk. The Nile is the child of snow; and its annual rise, on which the life of Egypt depends, is occasioned by its melting. The sun, approaching the summer solstice, finds the snows of the winter all treasured up for his magic touch to transform to water.

The snow tempers the heat of the atmosphere. And in this it has two opposite powers and offices. It heats and it cools. By being a non-conductor of heat, and at the same time translucent, it enables the Esquimaux to build his winter house entirely of its solid blocks; being the warmest substance for this purpose, in nature, so far as waste of heat is concerned, and at the same time serving the purpose of windows by transmitting the light in a broad mass into his humble dwelling. Animals live under its shelter in the severest cold. And the very earth is protected by it. Tender roots lie sheltered from the frosts by its thick covering. The winds that visit the south of India, our latitude, and Southern Europe, in summer, come across over the great snow tracts of the Himalaya, or the Alps, or the Rocky Mountains, charged with refreshing coolness.

And while its pure whiteness makes an agreeable change from the verdure of summer, it is admirably adapted to the various latitudes of the earth in modifying the light. The farther the sun withdraws from any part of the earth, the less light he emits there. And the snow follows him at respectful distance, increasing its bounty, as his rays diminish. The consequence is, our long winter nights are cheered by this brilliant covering that gathers and reflects all the scattered beams the sun has left. The long polar nights are not only illuminated, but also beautified by its wonderful phenomena.

Snow furnishes the most splendid material for road making. To it we owe the cheerful movement of the sleigh; and the lumbermen of our forests can do nothing until it has come to enable them to bring their timber to the streams. **Kirk.**

But few of us at home can realize the protecting power of this warm coverlet of snow. No eider-down in the cradle of an infant is tucked in more kindly than the sleeping dress of Winter about this feeble flower-life. The first warm snows, falling on a thickly-pleached carpet of grasses, heaths, and willows, enshrine the flowery growths which nestle round them in a non-conducting air-chamber; and, as each successive snow increases the thickness of the cover, we have, before the intense cold of winter sets in, a light, cellular bed covered by drift, six, eight or ten feet deep, in which the plants retain their vitality.

The early spring and late fall and summer snows are more cellular and less condensed than the nearly impalpable powder of winter. The drifts, therefore, that accumulate during nine months of the year, are dispersed in well-defined layers of different density. We have first the warm cellular snows which surround the plant; next, the fine, impacted snow-dust of winter; and above these, the later, humid deposits of the spring.

It is interesting to observe the effects of this disposition of layers upon the safety of the vegetable growths below them. These, at least in the earliest summer, occupy the inclined slopes that face the sun, and the several strata of snow take, of course, the same inclination. The consequence is, that as the upper snow is dissipated by the early thawings, and sinks upon the more compact layer below, it is, to a great extent, arrested, and runs off like rain from a slope of clay. The plant reposes thus in its cellular bed, guarded from the rush of waters, and protected too from the nightly frosts by the icy roof above it.

Dr. Kane.

There is a pretty, curious old town in Germany. The streets are narrow and the houses very quaint, with their pointed, gable-ends toward the street. One house stands somewhat isolated from the rest. It is at an angle where two streets meet, and is built with so many projections and jutting windows and carved friezes that it is quite a study.

One cold, cold afternoon in midwinter, when the silent frost was penetrating every where, and men moved quickly, muffled up in furs,—a time for people to close their doors, and gather round their firesides, —all the quiet inhabitants were astir. There was a bustle of preparation in parlor and kitchen; and young and old, wrapping their garments about them, were ready to go out in the cold. There were dismay and confusion in all the streets. Why?

They had heard that the French regiment, called the Pitiless, on its retreat from Moscow, was only three leagues off, and was to quarter in their village that night. There was every thing to fear from the revelry and excesses of soldiers, who acknowledged no right but that of the strongest.

In the queer old house of which we have spoken, there was no bustle of preparation. By the fire, in a large old room, sat an aged woman and her two grandchildren. Unable from her lameness to leave home, her grandchildren would not forsake her. Her faith in God enabled her to feel that they might be safer there than when fleeing from danger.

"O God! till darkness goeth hence,
Be thou our stay and our defense;
A wall, when foes oppress us sore,
To save and guard us evermore!"

These, the last notes of their evening hymn, died away amid the rafters of the shadowy room.

"Alas!" said the boy, mournfully, "we have no wall about us to-night to protect us from enemies."

"He will be our Wall himself," said the aged woman, reverently. "Think you his arm is shortened?"

"No, grandmother; but the thing is impossible without a miracle."

"Take care, my boy; nothing is impossible with God. Hath he not said he will be a wall of fire unto his people? We must trust him and he will be our wall of defense."

They sat quietly by the fireside. The wind moaned down the large open chimney, and the snow fell softly against the window-pane. Steadily it fell all night, and the wind drifted it in high banks, covering the shed, streets, walls, and paths of the silent and deserted town. And yet there was peace by that quiet fireside, the peace that can only be felt by the mind that is stayed on God. Few words were spoken. They held one another's hands, and looked into the fire, and listened, in the pauses of the storm, to catch the blast of the French trumpets. At nine o'clock the sound was faintly borne to them on the breeze; a few hurried blasts swept past them, intermingled with sounds of trampling feet and loud voices, and all was still.

Their hearts beat almost audibly; and they drew closer together, as they felt that they were now in the midst of their enemies. Helpless age and defenseless youth! What armor had they wherein to trust? The shield of faith! and safely they rested beneath its shadow!

Every house was a scene of revelry. Great fires were kindled. Altars were ransacked. The soldiers, with their songs and wine-cups, their oaths and blasphemy, made the streets ring, striving to drown the remembrance of intense cold and terrible privation in those hours of drunken merriment.

Still the little group in the quaint old house sat peacefully through the long, long hours of the night, till morning dawned and showed them the wall of defense which God had built round about them. Exposed as was their house, from its position, to the eddies and currents of the wind, the snow had so drifted about them that the doors and windows were completely blocked up; and the French soldiers had not found it. With the daylight they had left the town.

Wind and storm had fulfilled God's word, and encircled those who put their trust in him with a wall that protected them from their enemies, — a wall, not of fire, but of snow.

Under the Snow.

ALL in your gleaming folds of white,
    And robe the mountain-crest with snow;
Make each dark vale and hillside bright,
    Till all the earth is fair below!

Let all the forest oaks be clad,—
　The birch, and shivering aspen-tree,
Nor fail to shroud the tender buds
　And roots of frail anemone!

Then fold them close and shield them well
　To guard against the winter's cold;
While fast yon snowy flakes descend
　And penetrate the frozen mold.

And we will sing, though wild the storm,—
　Sing in December as in May,
Fill pledges from the icy brook,
　And all the year be glad and gay.

For when the winter solstice brings
　The Earth resplendent, like a bride,
She, with her snowy coronal,
　Fulfills her promise, glorified!

Fulfills her promise, when the days
　And months have brought the circling year,
When vernal woods and bursting buds
　And blue-eyed violets appear.

Faint type of what our hearts foretell,
　Of something glorious yet to be;
The life for which our spirits thirst,—
　And hid from all eternity!

And most for thee, high Hope, and Faith,
   I languish. Oh, return to me!
Come, fold me in thy heavenly robes,
   And mantle of sweet Charity.

And hide me from the outer world,
   And quicken to my inward sense
The earnest of those heavenly joys,
   Till I, departing, shall go hence.

And through those clouded days and brief,
   To yonder heaven I lift my eyes;
While, through life's frost and over-growth,
   The violet of my heart shall rise,

Up to that flaming Soul of Love,
   Who makes with joy my soul to sing;
And folds beneath the wintry snow
   The buds and garniture of spring.

S. D. C.

Wild-Flowers.

WE swept away the old sear leaves
   From off the wintry ground,
And underneath the frozen snow
   The tiny buds we found,

Just waiting for the beckoning nod
   Of sun and mellow air,
To spring in beauty from the sod,
   And shed their fragrance there.

So under many a frozen heart,
   Where Hope lies chilled and dead,
Beneath the tempest cares of life,
   Love's precious buds lie hid;
The smile of tender sympathy,
   The word of kindly cheer,
Like the warm, sunny air of spring,
   Will make the flowers appear.

                                       H. E. B.

The Alpine Violet.

MID Alpine wilds and thick-descending snows,
   Where massive hights uplift their towering crest,
Adown whose rugged sides no verdure glows,
   And e'en a feathery flake could scarcely rest;
Fearless alike of winter's rudest shock,
   Impending glaciers, or the tempest wild,
On the stern side of that cloud-piercing rock,
   A single violet looked up and smiled.

Some might have said, 'twas "born to blush unseen,
　　And waste its sweetness on the desert air;"
But no! Heaven's light shone in its humble mien,
　　And He was pleased who placed the floweret there.

So I, a human flower, am bid to grow,
　　Mayhap, in life's most wild, sequestered spot,
Where Sorrow's rudest tempests fiercely blow,
　　And woes accumulate o'ershade my lot.
Yet even here, alone and sadly chilled,
　　Nor known, nor noticed by the passing gaze,
Like that lone violet, I too may yield
　　To Heaven a silent offering of praise.
Earth may not know that I have ever been;
　　Yet pleasure to the eye of Heaven to give,
And thence, one sweet, approving smile to win,
　　Is, sure, no worthless mission to achieve.

　　　　　　　　　　　　　　　　Ib.

Instruction.

ANUARY! Darkness and light reign alike. Snow is on the frozen ground. Cold is in the air. The winter is blossoming in frost-flowers. Why is the ground hidden? Why is the earth white? So hath God wiped out the past: so hath he spread the earth, like an unwritten page, for a new year! Old sounds are silent in the forest and in the air. Insects are dead, birds are gone, leaves have perished, and all the foundations of soil remain. Upon this lies, white and tranquil, the emblem of newness and purity, the virgin robes of the yet unstained year!

April! The singing month. Many voices of many birds call for resurrection over the graves of flowers, and they come forth. Go, see what they have lost. What have ice and snow and storm done unto them? How did they fall into the earth stripped and bare? How do they come forth opening and glorified? Is it then so fearful a thing to lie in the grave?

In its wild career, shaking and scourged of storms through its orbit, the earth has scattered away no treasures. The hand that governs in April governed in January. You have not lost what God has only hidden. You lose nothing in struggle, in trial, in bitter distress. If called to shed thy joys as trees shed their leaves; if the affections be driven back into the heart, as the life of flowers to their roots, yet be patient. Thou shalt lift up thy leaf-covered boughs again. Thou shalt shoot forth from thy roots new flowers. Be patient! Wait! Beecher.

The sun had gone down before we entered the valley of Chamouni; the sky behind the mountain was clear, and it seemed for a few moments as if darkness was rapidly coming on. On our right hand were black, jagged, furrowed walls of mountain, and on our left Mont Blanc, with his fields of glaciers and worlds of snow; they seemed to hem us in and almost press us down. But in a few moments commenced a scene of transfiguration, more glorious than any thing I had witnessed yet. The cold, white, dismal fields gradually changed into hues of the most beautiful rose-color. A bank of white clouds, which rested above the mountains, kindled and glowed, as if some spirit of light had entered into them. You did not lose your idea of the dazzling, spiritual whiteness of the snow, yet you seemed to see it through a rosy vail. The sharp edges of the glaciers and the hollows between the peaks reflected wavering tints of lilac and purple. The effect was solemn and spiritual beyond any thing I had ever seen. These words, which had often been in my mind during the day, and which occurred to me more often than any others while I was traveling through the Alps, came into my mind with a pomp and magnificence of meaning unknown before:—"For by Him were all things created that are in heaven and that are in earth, visible and invisible, whether they be thrones, or dominions, or principalities, or powers; all things were created by him and for him; and he is before all things, and by him all things consist."

In this dazzling revelation I saw not that cold, distant, unfeeling fate, or that crushing regularity of wisdom and power, which was all the ancient Greek or modern deist can behold in God; but I beheld, as it were, crowned and glorified, one who had loved with our loves, and suffered with our sufferings. Those shining snows were as his garments on the Mount of Transfiguration, and that serene and ineffable atmosphere of tenderness and beauty, which seemed to change these dreary deserts into worlds of heavenly light, was to me an image of the light shed by his eternal love on the sins and sorrows of time, and the dread abyss of eternity.   **Sunny Memories.**

It is painful to think how our youth are coming to maturity without looking into these Treasure Houses of the King. The Bible and nature are mutually illustrative. And he has not a full Christian nature, who can not profoundly read and intensely enjoy both. As one has said of the book of nature,—and it might equally have been said of the Book of grace; —"The casual and general observer of nature soon ceases to be interested, because he looks only at the surface, and soon exhausts all the novelties. He merely stands on the outside of the temple, and, after gazing for a time at its noble proportions and splendid columns, his interest subsides. But he who really studies the works of God because he loves them is admitted into the penetralia, and there ten thousand new objects reward his search, opening continually before him, until he reaches the very Holy of Holies, and becomes a consecrated Priest." Education, then, consists in forming rather than in informing. Its two chief instruments are, not the wisdom of men, nor classical literature, but God's works and God's Book. Science without Revelation is of doubtful value. Revelation without Science is not seen in its fullness.

Let the snow-flake preach Sinai and Calvary, Eden and Gethsemane, the first and the second Adam. In its purity and beauty, its freshness and its bounty, it shows what a world this was when God made it. But when we see it rushing in the turbid stream, rolling in the dreadful avalanche, to bury the homes and possessions of men with their owners, — when we see it sullied and stained,—let us learn that man is fallen, that a curse is upon the earth and its lord, that "the whole creation groaneth and travaileth in pain together." **Kirk.**

God in Nature.

WHAT prodigies can Power divine perform
　　More grand than it produces year by year,
　　And all in sight of inattentive man!
　　Familiar with the effect, we slight the cause,
And in the constancy of Nature's course,
And regular return of genial months,
And renovation of a faded world,
See naught to wonder at.　Should God again,
As once in Gibeon, interrupt the race
Of the undeviating and punctual sun,
How would the world admire!　But speaks it less
An agency divine to make him know
His moment when to sink and when to rise,
Age after age, than to arrest his course?
All we behold is miracle; but seen
So duly, all is miracle in vain.

　　All this uniform uncolored scene
Shall be dismantled of its fleecy load
And flush into variety again.
From dearth to plenty, and from death to life,
Is Nature's progress when she lectures man

In heavenly truth; evincing, as she makes
The grand transition, that there lives and works
A soul in all things, and that soul is GOD.
The beauties of the wilderness are his,
That make so gay the solitary place
Where no eye sees them; and the fairer forms
That cultivation glories in are his.
He sets the bright procession on its way,
And marshals all the order of the year:
He marks the bounds which Winter may not pass
And blunts its pointed fury; in its case,
Russet and rude, folds up the tender germ,
Uninjured, with inimitable art;
And, ere one flowering season fades and dies,
Designs the blooming wonder of the next.

Cowper.

Teachings of Nature.

LOOK on this beautiful World and read the truth
    In her fair page; see, every season brings
New change to her of everlasting youth;
    Still the green soil with joyous living things
Swarms; the wide air is full of joyous wings;

And myriads still are happy in the sleep
  Of Ocean's azure gulfs, and where he flings
The restless surge. Eternal love doth keep,
In his complacent arms, the earth, the air, the deep.

Will, then, the Merciful One, who stamped our race
  With his own image, and who gave them sway
O'er earth, and the glad dwellers on her face,
  Now that our swarming nations far away
  Are spread, where'er the moist earth drinks the day,
Forget the ancient care that taught and nursed
  His latest offspring? Will he quench the ray,
Infused by his own forming smile at first,
And leave a work so fair all blighted and accursed?

Oh, no! a thousand cheerful omens give
  Hope of yet happier days, whose dawn is nigh.
He who has tamed the elements shall not live
  The slave of his own passions; he whose eye
  Unwinds the eternal dances of the sky,
And in the abyss of brightness dares to span
  The sun's broad circle, rising yet more high,—
In God's magnificent works his will shall scan,
And love and peace shall make their paradise with man.

W. C. Bryant.

CPSIA information can be obtained
at www.ICGtesting.com
Printed in the USA
BVHW062135150323
660508BV00004B/550